the S C I E N C E *library*

DISCOVERING SCIENCE

the SCIENCE *library*

DISCOVERING
SCIENCE

John Farndon and Ian Graham
Consultant: Robert Birke

Miles
Kelly
PUBLISHING

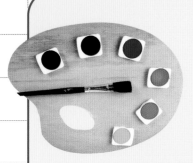

First published in 2004 by Miles Kelly Publishing Ltd
Bardfield Centre Great Bardfield Essex CM7 4SL

Copyright © 2004 Miles Kelly Publishing Ltd

This edition published in 2008

2 4 6 8 10 9 7 5 3

British Library Cataloguing-in-Publication Data
A catalogue record for this book is available from the British Library

Editorial Director Belinda Gallagher
Art Director Jo Brewer
Editor Jenni Rainford
Editorial Assistant Chloe Schroeter
Cover Design Simon Lee
Design Concept Debbie Meekcoms
Design Stonecastle Graphics
Consultant Robert Birke
Indexer Hilary Bird
Reprographics Stephan Davis, Ian Paulyn
Production Manager Elizabeth Brunwin

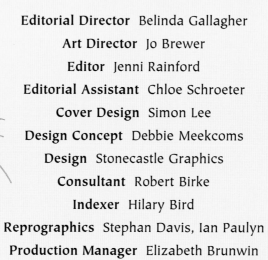

ISBN 978-1-84236-987-6

Printed in China

www.mileskelly.net
info@mileskelly.net

www.factsforprojects.com

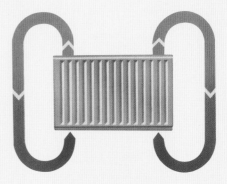

Contents

Solids, liquids and gases 8–9

The tiniest bits of all 10–11

Key substances 12–13

Chemicals and materials 14–15

Carbon chemicals 16–17

Electricity and magnetism 18–19

Electromagnetic radiation.. 20–21

Force and motion 22–23

Energy and work.............. 24–25

Heat......................... 26–27

Light 28–29

Sound 30–31

Air and water.................................. 32–33

Time 34–35

Glossary... 36–37

Index 38–40

How to use this book

DISCOVERING SCIENCE is packed with information, colour photos, diagrams, illustrations and features to help you learn more about science. Do you know what the world around us is made of or what makes things hot or cold? Did you know that atoms dance or that time is the fourth dimension? Enter the fascinating world of science and learn about why things happen, where things come from and how things work. Find out how to use this book and start your journey of scientific discovery.

Main text
Each page begins with an introduction to the different subject areas.

It's a fact
Key statistics and extra facts on each subject provide additional information.

Main image
Each topic is clearly illustrated. Some images are labelled, providing further information.

Check it out!
Find out more by surfing the Internet.

24

Energy and work

ENERGY IS the ability to make something happen. It is not just the light that comes from the Sun, or the heat that comes from a fire. Scientists say that energy is the capacity to do work. It is involved in everything that happens everywhere in the universe, however tiny or gigantic, from grass growing to stars exploding. All substances have energy locked up inside their atoms and molecules. Energy comes in many different forms and it can change from one form to another.

BRIGHT SPARKS

• Energy can never be created or destroyed; it can only be changed from one form to another. So all the energy in the universe has always existed no matter what form it is in now.

• Scientists define work as a force multiplied by the distance a load moves when the force acts on it. In the metric system, the unit of energy is called the joule. One joule is the work done when a force of 1 newton moves an object a distance of 1 metre, or a metre-newton. The foot-pound is the imperial unit used in countries such as the US.

IT'S A FACT

• All the energy in the universe may eventually turn to heat. This is called the 'Heat Death of the Universe' theory.

• Walking needs about five times as much energy as sitting still; running needs about seven times more energy.

Movement energy
A moving object has a type of energy called kinetic energy. The more massive the object is and the faster it moves, the more kinetic energy it h... When runners set off at the beginning of a race, they convert chemical energy in their muscles to kinetic energy. The faster they can change ch... energy into kinetic energy, the faster they run. At the end of the race th... stop producing kinetic energy. Air resistance and friction between thei... and the ground slow them down.

Read further > chemical energy
pg14 (d2)

→ At the end of the race, the runners quickly lose their kinetic energy and come to a halt.

← For a split second as the sprinters come away from the starting blocks, they accelerate faster than an average sports car.

Check it out!
• http://www.eia.doe.gov/kids/
• http://www.energyquest.ca.gov

A hot cup of coffee has about 40 per cent more energy than a tennis ball served at 25

1 2 3 4 5 6 7 8 9 10 11 12 13 14 15 16

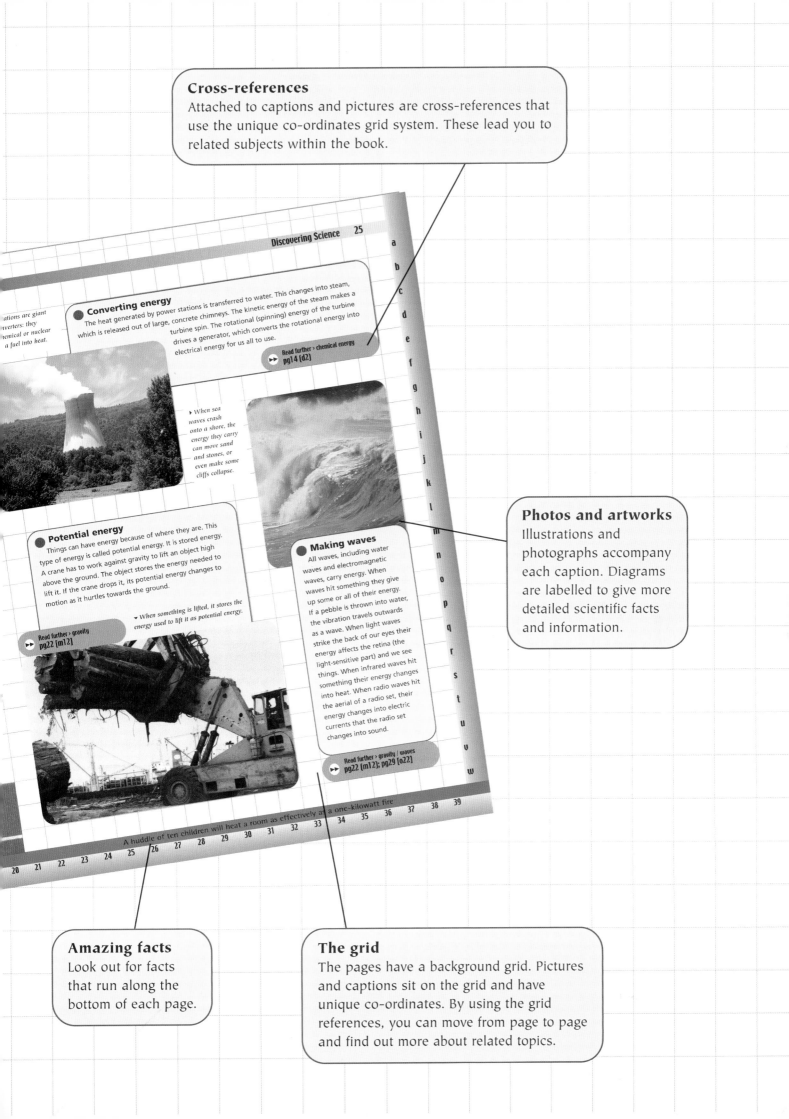

Cross-references
Attached to captions and pictures are cross-references that use the unique co-ordinates grid system. These lead you to related subjects within the book.

a
b
c
d
e
f
g
h
i
j
k
l
m
n
o
p
q
r
s
t
u
v
w

Converting energy
The heat generated by power stations is transferred to water. This changes into steam, which is released out of large, concrete chimneys. The kinetic energy of the steam makes a turbine spin. The rotational (spinning) energy of the turbine drives a generator, which converts the rotational energy into electrical energy for us all to use.

Read further › chemical energy
pg14 [d2]

...ations are giant ...verters: they ...hemical or nuclear a fuel into heat.

▶ When sea waves crash onto a shore, the energy they carry can move sand and stones, or even make some cliffs collapse.

Potential energy
Things can have energy because of where they are. This type of energy is called potential energy. It is stored energy. A crane has to work against gravity to lift an object high above the ground. The object stores the energy needed to lift it. If the crane drops it, its potential energy changes to motion as it hurtles towards the ground.

▼ When something is lifted, it stores the energy used to lift it as potential energy.

Read further › gravity
pg22 [m12]

Making waves
All waves, including water waves and electromagnetic waves, carry energy. When waves hit something they give up some or all of their energy. If a pebble is thrown into water, the vibration travels outwards as a wave. When light waves strike the back of our eyes their energy affects the retina (the light-sensitive part) and we see things. When infrared waves hit something their energy changes into heat. When radio waves hit the aerial of a radio set, their energy changes into electric currents that the radio set changes into sound.

Read further › gravity / waves
pg22 [m12]; pg29 [o22]

A huddle of ten children will heat a room as effectively as a one-kilowatt fire

20 21 22 23 24 25 26 27 28 29 30 31 32 33 34 35 36 37 38 39

Photos and artworks
Illustrations and photographs accompany each caption. Diagrams are labelled to give more detailed scientific facts and information.

Amazing facts
Look out for facts that run along the bottom of each page.

The grid
The pages have a background grid. Pictures and captions sit on the grid and have unique co-ordinates. By using the grid references, you can move from page to page and find out more about related topics.

Solids, liquids and gases

NEARLY EVERY substance in the universe is either a solid, a liquid or a gas. These are called states of matter. For example, rock is solid, water is liquid and oxygen is a gas. A substance can change from one state to another by gaining or losing energy. Heating water gives it more energy. The extra energy makes the water particles move about faster. If they have enough energy, they can escape from the water and form a gas – steam.

● IT'S A FACT

- The melting point of the metal, tungsten, is 3410°C. Its boiling point is 5900°C – about as hot as the surface of the Sun.

- The boiling point of the gas, helium, is −268.9°C!

● BRIGHT SPARKS

- The lowest possible temperature is −273°C, known as absolute zero, when molecules stop moving altogether.

- There is a fourth, less common state of matter besides solids, liquids and gases. This is plasma which is a bit like a gas but full of charged particles.

Some of the substances in lava (liquid rock) are heated so much that they change to gas

Warm air rises and cools

Gas (atoms or molecules) can move fast. They spread out to fill the space they are in

Water (liquid) cools and freezes as it becomes ice (solid)

Gas

Liquid

Liquid (atoms or molecules) can move about enough to flow past each other

Solid (atoms or molecules) stay in the same positions

● Rock, air and water

Solids, liquids and gases are everywhere in the world. The land is made of solid materials, such as rock and earth. Oceans and rivers are water, which is liquid. The air consists of many different gases *(see pg33 [h25])*. They may seem fixed, but they can all alter their state with a change in temperature or pressure.

▶▶ Read further › air / water / gases pg32 [m11]; pg33 [b22; b34]

Solid

🌐 Check it out!

- http://www.chem4kids.com/files/matter_states.html

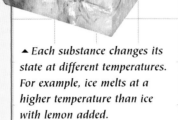

● So solid

Nearly every substance is made of tiny particles called molecules – too small to see with the eye alone. Solids have strength and a definite shape. The molecules are firmly bonded together in a regular framework. All molecules move non-stop, but in a solid they simply vibrate on the spot. The hotter the solid becomes, the more they vibrate. If it gets hot enough the molecules vibrate so much that the framework breaks down and the solid melts, such as ice turning to water.

▲ *Each substance changes its state at different temperatures. For example, ice melts at a higher temperature than ice with lemon added.*

►► **Read further › molecules**
pg10 (j15); pg26 (n2)

● Boiling and melting

The temperature at which a substance melts from a solid to a liquid is called its melting point. The highest temperature a liquid can reach before turning to a gas is called its boiling point – although some of the liquid may evaporate (turn to gas) before it reaches this point. Each substance, such as water or chocolate, has its own melting and boiling point. Water melts at 0°C and boils at 100°C. When a gas cools down enough, it condenses to a liquid, such as when steam turns to water. When a liquid cools down enough, it turns solid or freezes, such as when water turns to ice.

►► **Read further › heat**
pg26 (d2)

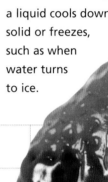

● Going liquid

Unlike solids, liquids such as water have no shape of their own and therefore flow into the shape of any container they are poured into. The molecules in liquids are partly bound together in clusters. Liquids flow because these bonds are loose enough for the molecules to move about. The clusters roll over each other like dry sand, allowing the liquid to flow freely and quickly.

▲ *Water, like all liquids, takes the shape of whatever it is poured into.*

▸ *The particles in solid chocolate gain energy as they are heated – breaking away from each other as the solid melts.*

►► **Read further › moving molecules**
pg10 (j15)

◂ *The molecules in water flow freely over each other allowing liquids to pour quickly, such as the water in this waterfall.*

● What a gas

Like liquids, gases have no shape or strength of their own. Unlike liquids, they have no fixed volume (mass or amount of space taken) either so they spread out to fill the space they are put in. In the same way, they can be squeezed into a much smaller space.

◂ *Airships float because the gas (helium) inside is lighter than the air outside.*

►► **Read further › gases / air**
pg13 (d29); pg31 (b22)

The word vapour is used to describe a gas that is usually found as a liquid, for example water

a b c d e f g h i j l m n o p q r s t u v w

The tiniest bits of all

MATTER IS every substance in the universe – everything that is not just empty space. Yet matter itself, even the most solid rock, is largely empty space. All matter is made of tiny particles, atoms, with empty space between them. Atoms, and the spaces between them, are far too small to see except with the most amazingly powerful microscopes. You could fit two billion atoms on this full stop. Even atoms are not solid, though. They are more like clouds of energy, dotted with tinier particles called sub-atomic particles.

● **IT'S A FACT**

• By smashing atoms together at great speeds, scientists have found over 200 sub-atomic particles, but few last for more than a fraction of a second.

• Among the tiniest particles of all are neutrinos. They are thousands of times lighter than electrons.

▶▶ **Read further › compounds / elements**
pg13 (b22; l29); pg15 (c22)

▶ *Particles with opposite electrical charges (positive and negative) attract each other. A proton has a positive electrical charge. An electron has an equal but opposite (negative) electric charge. Atoms contain protons and electrons, which attract each other, holding the whole atom together.*

▲ *In a hydrogen atom, one electron whizzes around one proton nucleus.*

▲ *In a helium atom, two electrons whizz around a nucleus of two protons and two neutrons.*

● **Atomic partners**

Atoms usually bond (join) together in groups to form combinations called molecules. A molecule is the smallest particle of a substance that can exist on its own. For example, a molecule of oxygen, the gas we breath in in order to live, occurs as a pair of oxygen atoms bonded together. Water, also essential for life, is a molecule of two atoms of hydrogen and one of oxygen joined together.

● **Atomic**

At the centre of an atom is a nucleus (dense cluster) of two kinds of particle: protons and neutrons. Even tinier particles, called electrons, zoom around the nucleus (see pg11 [q33]). The various sub-atomic particles are just concentrations of energy that are likely to occur in certain places. Protons have a positive electrical charge, electrons have a negative electrical charge and neutrons have no charge.

● Electron
● Proton
● Neutron

▶ *In an oxygen atom, eight electrons whizz around a nucleus of eight protons and eight neutrons.*

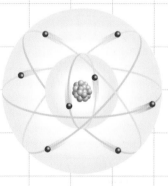

▼ *A carbon dioxide molecule (waste product when we breathe out) is a chemical compound that consists of one carbon atom and two oxygen atoms. Carbon dioxide has the chemical formula CO_2.*

▶▶ **Read further › electricity**
pg18 (l9); pg19 (e22)

 Check it out!

• http://www.chem4kids.com/files/atom_intro.html
• http://www.pbs.org/wgbh/aso/tryit/atom/

If an atom were the size of a sports arena, its nucleus would be the size of a pea!

Crystal gazing

Most naturally occurring solids form crystals. Crystals are hard, shiny chunks that form in uniform geometric shapes. Each crystal is made from a regular framework or lattice of atoms or molecules. Grains of sugar *(see pg33 [r28])* and salt are crystals. So, too, are most gems, such as diamonds and emeralds. Most rocks and metals are built up from crystals, too, although the individual crystals are often far too small to see with the naked eye.

▲ *Diamond, made of carbon atoms linked together in a rigid structure, is the hardest natural material.*

Read further > metals / carbon
pg14 (i14); pg16 (i13)

BRIGHT SPARKS

• Most of an atom is empty space. The distance from its nucleus to the closest electron is about 5000 times the size of the nucleus. If the nucleus was 1 cm wide, the closest electron would be about 50 m away.

• Protons normally push each other away because they are all positively charged. Inside an atom a powerful force, called a strong nuclear force, holds them togehter and stops the nucleus from flying apart.

Different atoms

Each of nature's 100 or so basic chemicals or elements is made up from an atom with a certain number of protons in its nucleus. A uranium atom has 92 protons, the most in any element found in large amounts in nature. In each atom, the number of protons is usually matched by the same number of electrons, which are arranged in rings or 'shells' around the nucleus. The way an atom reacts with other atoms (its chemical character) depends on how many electrons there are in its outer shell.

Read further > electron shells
pg12 (d2; h9)

● *Neutron*

● *Proton*

○ *Electron*

▸ *In the centre of the atom is the nucleus, which is made up of equal numbers of protons and neutrons. These are held together by a very strong force, which can be used to create nuclear energy.*

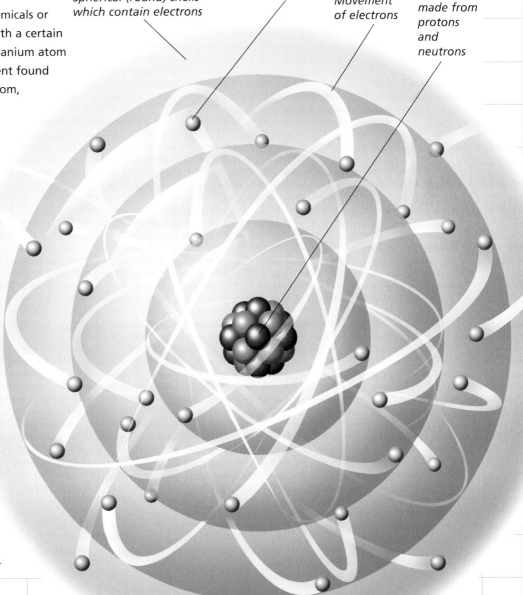

Spherical (round) shells which contain electrons

Electrons

Movement of electrons

Nucleus made from protons and neutrons

The number 10 followed by 80 zeros is the number of atoms in the universe

Key substances

A LL THE substances in the universe can be split into smaller pieces until they become the simplest known substances, chemical elements. Examples are gold, carbon, and oxygen. Each element has unique chemical and physical features because it is made up from its own kind of atoms. All the atoms of an element are the same, with the same number of protons and other parts (see pg11 [n33]), and different from the atoms of all other elements.

Period or row 1A is a 'special case' with just the two lightest elements: hydrogen and helium

▼ *Elements also change their characteristics out of step with the Groups/Periods in two subsets: A and B.*

● Arranging the elements

An element's atomic number is the number of protons in its nucleus. Elements can be listed from the lightest, hydrogen (atomic number 1), to heavier elements, such as lawrencium (atomic number 103). Russian chemist Dimitri Mendeleév arranged them in the Periodic Table. The elements in a vertical column, or Group, are similar to each other in chemical and physical features, but get heavier in 'jumps' downwards. The elements in a horizontal row, or Period, are heavier from left to right by one each time. Each also becomes less reactive or able to join with other elements, depending on the number of electrons and gaps for electrons in its outermost shell (see pg11 [k31]]). The most reactive elements are on the left side, and the least reactive on the right side (see pg13 [k26]).

Group or column 2A elements (green), called alkaline-earth metals, are fairly soft and reactive

Elements in Groups 3B onwards (purple) are mostly hard and shiny, known as transition metals

▶▶ Read further > atoms / metals
pg11 (i22); pg15 (c22)

H Hydrogen 1													
Li Lithium 3	Be Beryllium 4											B Boron 5	C Carbon 6
Na Sodium 11	Mg Magnesium 12											Al Aluminium 13	Si Silicon 14
K Potassium 19	Ca Calcium 20	Sc Scandium 21	Ti Titanium 22	V Vanadium 23	Cr Chromium 24	Mn Manganese 25	Fe Iron 26	Co Cobalt 27	Ni Nickel 28	Cu Copper 29	Zn Zinc 30	Ga Gallium 31	Ge Germaniu 32
Rb Rubidium 37	Sr Strontium 38	Y Yttrium 39	Zr Zirconium 40	Nb Niobium 41	Mo Molybdenum 42	Tc Technetium 43	Ru Ruthenium 44	Rh Rhodium 45	Pd Palladium 46	Ag Silver 47	Cd Cadmium 48	In Indium 49	Sn Tin 50
Cs Caesium 55	Ba Barium 56		Hf Hafnium 72	Ta Tantalum 73	W Tungsten 74	Re Rhenium 75	Os Osmium 76	Ir Iridium 77	Pt Platanium 78	Au Gold 79	Hg Mercury 80	Tl Thalium 81	Pb Lead 82
Fr Francium 87	Ra Radium 88		Rf Rutherfordium 104	Db Dubnium 105	Sg Seaborgium 106	Bh Bohrium 107	Hs Hassium 108	Mt Meitnerium 109	Uun Ununnilium 110	Uuu Unununium 111	Uub Ununbium 112		
			La Lanthanum 57	Ce Cerium 58	Pr Praseodymium 59	Nd Neodymium 60	Pm Promethium 61	Sm Samarium 62	Eu Europium 63	Gd Gadoloinium 64	Tb Terbium 65	Dy Dysprosium 66	Ho Holmiu 67
			Ac Actinium 89	Th Thorium 90	Pa Protactinium 91	U Uranium 92	Np Neptunium 93	Pu Plutonium 94	Am Americium 95	Cm Curium 96	Bk Berkelium 97	Cf Californium 98	Es Einsteiniu 99

Group 1 elements (blue) are soft and react readily, known as alkali metals

These two rows, (orange and green) lanthanides and actinides, each fit into one position in the main table

Since 1969, more than 30 elements with large atoms, called 'transuranic' elements, have been discovered

| 1 | 2 | 3 | 4 | 5 | 6 | 7 | 8 | 9 | 10 | 11 | 12 | 13 | 14 | 15 | 16 | 17 | 18 | 19 |

BRIGHT SPARKS

- In addition to the atomic number, each element also has an atomic mass. This is the relative 'weight' of the atom's whole nucleus – the protons and neutrons added together. Lead's atomic number is 82 but its atomic mass is 207.

- Scientists give each element a symbol. This is usually the first letter of its name, like O for oxygen and C for carbon. If two or more elements begin with the same letter, a second small letter may be added. So hydrogen is H and helium is He.

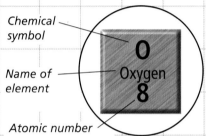

Chemical symbol

Name of element

O
Oxygen
8

Atomic number indicating number of protons in atom's nucleus

Gases most noble

The Group 8 elements down the furthest right of the Periodic Table is a very special Group. It is also called Group 8 or 0, because the atoms of these elements have no electrons missing from their outer shells. With a full outer shell of electrons, their atoms have no need to share electrons with other atoms. They are therefore extremely stable and unreactive. They are also called the noble gases, because they stay noble (apart from) other chemicals. Noble gases such as argon and krypton are used in light bulbs just because they are so unreactive, so will not burn out the filament – the tiny, thin coiled wire inside the bulb. Neon is used to make neon lights for the same reason, so it can burn brightly without reacting.

Read further > gases
pg9 (q31); pg33 (b22)

▲ An electric current sent to the lightbulb makes the filament glow to create light. Argon inside the bulb saves the filament from burning out.

He	Group 8 elements (light blue) are known as noble gases and hardly ever react with other elements
Helium	
2	

N	O	F	Ne
Nitrogen	Oxygen	Flourine	Neon
7	8	9	10

P	S	Cl	Ar
hosphorus	Sulphur	Chlorine	Argon
15	16	17	18

As	Se	Br	Kr
Arsenic	Selentium	Bromine	Krypton
33	34	35	36

Sb	Te	I	Xe
Antimony	Tellurium	Iodine	Xenon
51	52	53	54

Bi	Po	At	Rn
Bismuth	Polonium	Astatine	Radon
83	84	85	86

ome elements in Groups A to 7A (green) are alled poor metals

The other elements in Groups 3A to 7A (brown) are called non-metals

Compounds

Pure elements are quite rare in the world. Most substances are made of two or more elements joined together in a compound. A compound is not just a mixture of the elements. When the elements combine they are changed chemically to make an entirely new substance. Sodium, for instance, is an element that fizzes violently when dropped in water, while chlorine is a thick, green gas, but they combine to make a compound called sodium chloride, which is ordinary table salt.

▲ Eggs are a compound of sulphur, carbon, nitrogen, phosphorus, hydrogen and oxygen.

▼ Citric acid, found in lemon juice, is a compound of hydrogen, oxygen and carbon mixed with water.

Read further > ores /salt
pg14 (i14); pg15 (k31)

Er	Tm	Yb	Lu
Erbium	Thulium	Ytterbium	Lutetium
68	69	70	71

Fm	Md	No	Lr
Fermium	Mendelevium	Nobelium	Lawrencium
100	101	102	103

Check it out!
- http://www.funbrain.com/ periodic/
- http://www.chem4kids.com/ files/elem_pertable.html

▶ When food products such as eggs, butter and sugar are mixed together and cooked, the heat bonds the different compounds together into a new compound.

a e f g h i j k l m n o p q r s t u v w

Chemicals and materials

THE UNIVERSE is made up from millions of different substances that we know about – and probably millions more that have not yet been discovered. Barely 100 of these are pure chemical elements consisting of identical atoms. Most are compounds made from different combinations of these atoms. Many natural substances, such as wood, soil and rocks, are a mixture of two or more compounds. Metals exist naturally as compounds and pure water is a compound of the elements hydrogen and oxygen. Tap water and sea water are mixtures, for there are always other substances mixed in with them. Some substances, when mixed with water, form an acid.

● IT'S A FACT

• The oldest known alloy is bronze, made at least 5000 years ago, by mixing copper and tin.

• Mercury is the only metal that is liquid at normal temperatures. It freezes when the temperature drops to -38.87°C.

● Mixed metals

Metals are very rarely pure. Most occur naturally in the ground in compounds called ores, and the metal must be extracted by heating and other processes. Even then, metals usually contain some impurities. Sometimes impurities are added deliberately to create an 'alloy', which gives the metal a particular quality, such as resistance to corrosion or extra strength. Carbon is added to iron to make an incredibly tough alloy called steel, and chromium is added to steel to make stainless steel, which does not rust or stain.

◀ *Alloys of aluminium and magnesium are very strong and corrosion-resistant. This makes them ideal for producing vehicle frames and buildings that have to withstand the weather and environmental pollution.*

▶▶ **Read further > carbon**
pg16 (d2; i13)

Gold is the only metal that does not rust

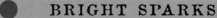

▸ *Metals such as steel are used in building tough structures such as cars.*

Metallic

Three out of every four elements is a metal, such as gold or iron. Most metals are shiny, hard substances that ring when you hit them. They are mostly tough, yet can often be easily shaped – either by hammering or by melting into moulds. This makes them wonderful materials for making everything from spoons to cars and space-rockets. The atom of metals knit together into a strong framework or lattice. Atoms within the lattice share their electrons freely. The freely dancing electrons make metals great conductors of heat and electricity, since the moving electrons pass them on like batons in a relay race.

Read further › conduction
pg27 (c22; r30)

▸ *Anybody can float easily in the Dead Sea in Israel. The very high salt content gives greater upward force than fresh water.*

Read further › hydrogen
pg12 (d2); pg33 (b22; b34)

Liquids that burn

When some substances dissolve in water they create a special liquid called an acid. A drink that tastes sour, such as lemon juice, is a weak acid. Strong acids, such as sulphuric acid, are highly corrosive, attacking clothes and skin, and dissolving metals. All acids, both weak and strong, contain hydrogen. When mixed with water, the hydrogen atoms lose their one electron, and become ions – electrically charged atoms. It is these ions that make acids sour tasting and corrosive.

▸ *The chemical opposite of an acid is a base. Strong bases, such as caustic soda, are so corrosive that they are dangerous so protective gloves and clothing is worn when they are handled. Weak bases, such as baking powder, taste bitter or may have a soapy feel.*

BRIGHT SPARKS

• Tap water often has traces of dissolved salts such as calcium carbonate, which make water 'hard'. Hard water can create limescale around taps and make soap slow to lather.

• Acids play a crucial role in the human body. Amino acids make protein. The acid DNA, found in cells inside the body, provides instructions for life.

Salt of the earth

Table salt is just one of many substances called salt. Many of the minerals that make up the rocks of the Earth's surface are salts. Salts are a special kind of solid made of crystals that can take a variety of shapes. Salt is obtained from sea water in hot countries. Some sea water has a high salt content. The water is held in shallow pools and the salt gathered when all the water has evaporated in the heat. A salt forms when an acid and a base react together. For example, table salt forms when the base sodium hydroxide reacts with hydrochloric acid. Most salts dissolve in water, which makes them very useful materials for living things. Salts in the body maintain water balance and keep nerve signals healthy.

Read further › crystals / dissolving
pg11 (b22); pg33 (m22)

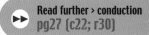

 Check it out!

• http://www.miamisci.org/ph/phpanel.html

The most abundant – commonly-found – metal in the world is aluminium

Carbon chemicals

CARBON IS a very special element. The hardest known substance, diamond, is carbon. So is coal and the graphite in pencils. Carbon has an ability to form compounds easily because of the structure of its atom. There are more than 1 million known carbon compounds, from limestone to diesel oil. With four out of the full eight electrons in its outer shell, a carbon atom can form compounds either by gaining electrons or by losing them. This means it will join up with just about anything to make a wide variety of products, from oil-based paints to parachutes.

● IT'S A FACT

• Most plastics are made from a group of chemicals called ethenes, which are derived from petroleum.

• Natural diamonds formed deep in the Earth billions of years ago.

● Carbon

Pure carbon occurs in four allotropes (forms): diamond, graphite, soot and charcoal and a special manufactured form called fullerene. Graphite can be stretched out into long fibres called carbon fibres.

◀ *When bonded together, carbon fibres make an incredibly light but tough material, ideal for making items, such as oars used for rowing.*

Diesel   *Graphite*

Carbon

Diamond   *Charcoal*

◀ *Carbon and its compounds can be used for many things: diesel to run vehicles; graphite in pencils; diamonds in jewellery; and charcoal burnt to produce energy.*

►► Read further > alloys
pg14 (i14)

Despite being present in so many things, carbon makes up just 0.032% of the Earth's crust

1   2   3   4   5   6   7   8   9   10   11   12   13   14   15   16   17   18   19

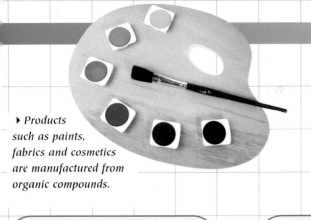

▸ Products such as paints, fabrics and cosmetics are manufactured from organic compounds.

▸ The bow shape of this parachute is created by the air pushing up against it, which helps to slow down the speed at which the forces of gravity pull the parachutist down to the ground.

Organic chemicals

Carbon has an almost unique ability not only to form compounds with other elements, but also to join together with other carbon atoms as well to form complex chains and rings. Complex carbon chain and ring molecules are the basic chemicals that life itself depends on. For example, the proteins from which the body is built are all carbon compounds. There is such a huge variety of carbon compounds that there is an entire branch of chemistry called organic chemistry devoted to their study.

▶▶ Read further › compounds
pg13 (l29)

Plastic world

Plastics are among the most amazing of all materials, used for everything from drinks bottles to car bodies. Light and easy to shape into any form, they can be made as soft as silk or as strong as steel. They are entirely manufactured: the secret is to get molecules of carbon compounds – mainly carbon and hydrogen – to link up in long chains called polymers. In some plastics, the chains are tangled together like spaghetti to make them strong yet flexible. These are ideal for making items such as parachutes, which need to be strong enough to support weight and yet flexible enough to glide in the air. Chains that are held rigidly together make stiff plastic, used for items such as window frames.

▶▶ Read further › molecules / steel
pg10 (i14); pg14 (i14)

▶▶ Read further › the Sun's energy
pg21 (b22)

The carbon cycle

Most carbon atoms have existed since the beginning of the Earth and, through a process called the carbon cycle, circulate through animals, plants and the air. The leaves and stems of every plant are built largely from a natural material called cellulose. Like plastic, cellulose is a polymer – a long chain of carbon-based molecules. Plants put these chains together from sugar molecules called glucose, which they make with water and carbon dioxide from the air using energy from the Sun. Animals eat plants, using the carbon compounds taken in.

▸ When living things decay, fuels burn and plants and animals break down sugars to release energy. Carbon dioxide is released into the air during this process, thus completing the carbon cycle.

BRIGHT SPARKS

• Plastic's durability makes it hard to dispose of, so scientists have developed kinds that can be degraded (broken down) by light or bacteria. However, the world still faces increasing problems with plastic waste.

• Fullerenes are molecules between 32 and 600 carbon atoms linked together in a ball shape. The first fullerene was a ball of 60 carbon atoms. It was named Buckminsterfullerene, or the 'Bucky Ball', after the architect R. Buckminster Fuller who made domes that the Bucky Ball looked like.

Carbon has the chemical formula C and atomic number 6

a b c d e f g h i j k l m n o p q r s t u v w

Electricity and magnetism

IT'S A FACT

• Lightning is the sudden release of a giant charge of static electricity that builds up inside storm clouds.

• The best conductors of electricity are materials that contain lots of free electrons, such as copper and silver.

ELECTRICITY IS one of the most useful of all forms of energy, providing us with everything from heat and light to the tiny pulses that make computers work. Electricity is closely linked with magnetism – the invisible force between magnetic materials. When electricity moves, magnetism is created. When magnets move, electricity is created. Together, electricity and magnetism form one of the forces that holds the universe together – electromagnetism.

▼ Pylons carry electric cables from power stations through a distribution grid. Electricity can travel safely, high above groun...

BRIGHT SPARKS

• The Earth is a giant magnet. If a magnet is left to move freely, the Earth's magnetic field ensures that it always points with one end aimed at the North Pole and the other at the South Pole.

• Materials that do not conduct electricity very well are called insulators. Plastic and rubber are good insulators.

● Electric currents

When a magnet and a coil of wire move near each other, the magnet induces (creates) an electric current in the wire by pushing electrons along it. Power stations use this to make electricity. Flowing water or a jet of steam spins coils of wire around powerful magnets. Electric currents are induced in the coils. The electricity then flows along wires to our homes, schools and workplaces. The wires are either held above ground on top of tall pylons or are buried underground.

High pylons hold electric cables safely above ground

Cables carry flo of electricity

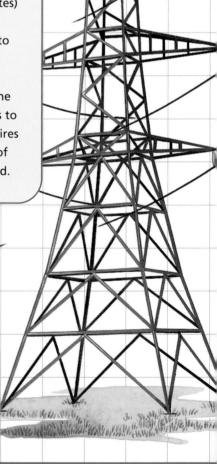

▼ Electricity is made from atoms that move along a wire.

▶▶ Read further > currents pg19 (m22)

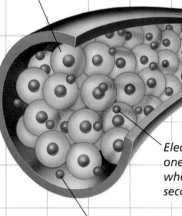

Atom

Electrons can be pushed from one atom to the next and when billions do this every second, electricity flows

Electron

● Check it out!

• http://www.mos.org/sln/toe/toe.html
• http://www.factmonster.com/ce6/ sci/A0831162.html

The first known magnets were not iron but stones that contained iron, called lodestones

▶ *Charged particles such as those generated by a Van de Graaf generator, can make hair stand on end. All the hairs have the same charge (all positive or all negative) and so repel (push apart) each other. They stand on end to get away from each other.*

Static electricity

Electricity is caused by the behaviour of electrons, which are electrically charged particles. If a material gains extra electrons it becomes negatively charged. If a material loses electrons it becomes positively charged. When two materials rub together, electrons may rub from one onto the other. They both become electrically charged, one positive and the other negative. This is called static electricity, because the electric charges on the charged material do not move – they are static (stationary).

Read further › protons / electrons
pg10 (o2); pg11 (i22)

Magnetic attraction

Magnets are special pieces of metal – usually iron – that have the power to attract magnetic materials such as iron and steel. Around each magnet is a region or 'field' where its effect is felt. The field is strongest at the two ends or 'poles' of the magnet, and gets weaker further away. The magnet's power works in opposite directions at each pole, so one is called a north pole and the other a south pole. While opposite poles on two magnets attract each other, identical poles repel (push each other apart). So north poles attract south poles but repel north poles.

Read further › iron / water
pg15 (c22); pg33 (b34)

▶ *Objects containing iron, such as nails or screwdrivers are affected by magnetic fields.*

Magnetic field

Magnetic lines of force

Electric currents

Electrons move easily through material called conductors. Metals, such as copper and gold, are used as connectors in electric circuits because they are very good electrical conductors. They have many electrons, which can move easily through wires. When lots of electrons move in the same direction an electric current is produced. These flow around loops, or circuits. Batteries provide the energy to drive these electrons around a closed loop. They produce currents that flow instantly in one direction – direct currrent (DC).

▲ *Power stations generate currents that reverse many times every second. Unlike the currents in batteries, power station currents are called alternating currents (AC).*

Electric magnets

When an electric current flows through a wire, it produces a magnetic field around the wire. The magnetic field is stronger when the wire is coiled around a piece of iron. This type of 'electric magnet' is called an electromagnet. Unlike a bar magnet, an electromagnet can be switched on and off. The magnetic field disappears when the electric current is switched off.

▶ *An electromagnet can lift a car into the air by attracting the iron-based steel of the car's body.*

Read further › electrical charges
pg10 (o2)

Wet skin conducts electricity up to 1000 times better than dry skin, but can cause dangerous electric shocks

a b c d e f g h i j k l m n o p q r s t u v w

Electromagnetic radiation

LIGHT BEAMS, radio signals, microwaves used for cooking, heat rays from fires and X-rays used in hospitals are all part of the same family of electromagnetic radiation. They are called electromagnetic waves because they are partly electric and partly magnetic. They travel in straight lines and at the same speed – the speed of light. In the vacuum of space, electromagnetic waves travel at 300 million metres per second. At that speed, they could travel around the world in about one-tenth of a second. The difference between electromagnetic waves is the length of each wave.

IT'S A FACT

• Glass lets visible light pass through, but not infra-red (heat) rays. Greenhouses are warm inside because light can enter and warm the contents, but the heat rays produced inside the greenhouse are trapped by glass.

• Nearly all the fuels we use today, including oil and wood, were produced by the action of electromagnetic radiation from the Sun on plants.

BRIGHT SPARKS

• Scientists can track the movements of all sorts of creatures, from whales and polar bears to tigers and elephants, by fitting them with radio collars. A radio collar transmits (sends out) a radio signal that can be received a long way away, even by satellite, allowing the animal's position to be marked on a map.

• Weather satellites can take pictures of the world's weather at night by using infra-red cameras.

Energy waves

Electrons emit a huge range or 'spectrum' of electromagnetic waves. The light we see, called visible light, is just a small portion in the middle of the spectrum. At one end of the spectrum are waves too long for our eyes to see, including radio waves and microwaves. At the other end are waves too short for us to see, including ultraviolet light and X-rays.

◄ *The varying lengths of electromagnetic waves are each useful for different purposes.*

▶▶ Read further › electromagnetism pg18 [f2]

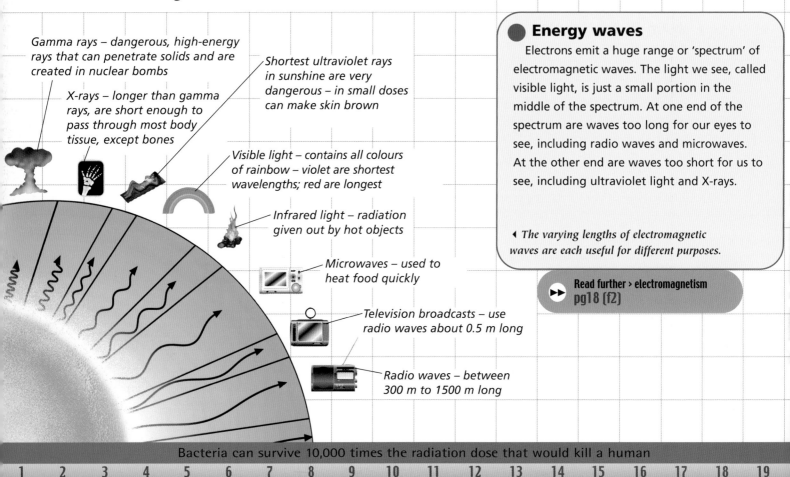

Gamma rays – dangerous, high-energy rays that can penetrate solids and are created in nuclear bombs

X-rays – longer than gamma rays, are short enough to pass through most body tissue, except bones

Shortest ultraviolet rays in sunshine are very dangerous – in small doses can make skin brown

Visible light – contains all colours of rainbow – violet are shortest wavelengths; red are longest

Infrared light – radiation given out by hot objects

Microwaves – used to heat food quickly

Television broadcasts – use radio waves about 0.5 m long

Radio waves – between 300 m to 1500 m long

Bacteria can survive 10,000 times the radiation dose that would kill a human

| 1 | 2 | 3 | 4 | 5 | 6 | 7 | 8 | 9 | 10 | 11 | 12 | 13 | 14 | 15 | 16 | 17 | 18 | 19 |

Sun

Most of the radiation hitting the Earth comes from the Sun, which pushes out huge amounts of energy. Some of the Sun's radiation is waves, such as light and X-rays. Fortunately, Earth's atmosphere only lets through the light and warmth that we need, and shields us from the most harmful waves, such as extreme ultraviolet and X-rays.

▶▶ **Read further › magnetic fields**
pg19 (b31)

▲ *The Earth's atmosphere is vital, allowing for humans and animals to breathe, and keeping out the most harmful radiation from the Sun.*

Radiation hazard

Some electromagnetic radiation can be dangerous. Even low energy radiation from the Sun can cause harmful diseases, such as skin cancer, after prolonged exposure from sunbathing. But it is short wave, high-energy rays, such as X-rays and gamma rays that are the main hazards. Rays like these damage living tissues by 'ionizing' the atoms in them – that is, knocking electrons off. This can cause damage to the biological mechanisms of the body. This is why people who work with X-rays in hospitals are protected behind screens.

◀ *Prolonged exposure to ultraviolet (UV) rays can cause serious damage. Suntan lotions containing a block against these UVA and UVB rays help to prevent sunburn and diseases such as skin cancer.*

▶▶ **Read further › atoms / electrons**
pg10 (i22)

Heat pictures

Hot objects emit (give out) electromagnetic waves. We are unable to see them, but thermal cameras can detect them and make pictures from them. In a thermal picture the hottest things are the brightest and the coldest things are the darkest. Thermal cameras can take pictures in total darkness because they do not rely on light. They can be useful in showing how animals behave in the wild at night, without having to shine bright lights on them.

▶ *Thermal cameras are used to to diagnose illnesses by highlighting different parts of the body according to temperature. The yellow areas show heat, and possibly disease, blue areas show cooler parts of the body.*

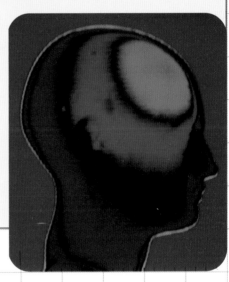

▶ *Satellites and spacecraft use radio waves to send pictures and other information to Earth.*

Space waves

Unlike sound waves, which need something to travel through, electromagnetic waves can travel through empty space. This property is very useful. It lets us see distant stars at night – the light from them has to travel through empty space to reach us. It also lets us speak to astronauts in space and communicate with satellites by radio.

▶▶ **Read further › waves**
pg25 (m34)

a b c d e f g h i j k l m n o p q r s t u v w

Force and motion

FORCES ARE pushes and pulls. They change the speed, direction or shape of things. Some forces act only when things touch each other, like kicking a football. Other forces, including gravity and magnetism, act at a distance. Forces always occur in pairs – a force acting in one direction always creates an equal force acting in the opposite direction. The two forces are called the action and the reaction. When you push against a wall, the wall pushes back equally hard. If it did not, your hand would go through it. The main types of forces in nature are gravitational (gravity), electrical, magnetic and nuclear.

▲ When sky-divers open their parachute, its air resistance force acts in the opposite direction to gravity, slowing them down.

● Rollercoaster ride

A rollercoaster has no motor. It gets its thrilling speed from gravity. It is towed up high and released. As it hurtles downhill, the constant pull of gravity makes it go faster and faster. When it reaches the bottom of a slope it is travelling so fast that it carries on uphill to the top of the next slope. The tendency of a moving body to keep moving is called momentum.

▶▶ **Read further › gravity**
pg23 (j28)

● Falling down forcefully

One of Newton's most startling discoveries was that things do not just fall of their own accord: they fall because they are pulled down by a force called gravity. Gravity is the force of attraction that pulls everything towards the centre of the Earth. Every single bit of matter in the universe, no matter how tiny, exerts its own gravitational pull on other matter. The strength of the pull depends on the mass (matter) of the object. More massive objects exert a stronger pull of gravity. The further apart things are, the weaker their pull of gravity on each other.

▶▶ **Read further › Newton / black holes**
pg23 (q30); pg35 (m22)

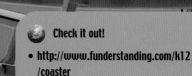

Particles of light called photons have almost no mass, so they can be accelerated faster than anything else

| 1 | 2 | 3 | 4 | 5 | 6 | 7 | 8 | 9 | 10 | 11 | 12 | 13 | 14 | 15 | 16 | 17 | 18 | 19 |

Ice dancer being lifted does 'work' by moving and holding her body in correct position

Work, effort and load

Work, effort and load are important concepts in physics, especially in relation to machines, which usually move something. Load describes the size of the object moved, measured in kilograms or pounds. Effort describes the force used to move it, measured in newtons or pounds. Work describes just how long the effort is applied for – or more specifically the force used times the distance moved by the load. In the metric system, the unit of work is the joule – the work done when a force of 1 newton moves something 1 metre. One joule equals 1 newton-metre. In the US, the unit of work is a foot-pound. This is the work done when a force of 1 lb (pound) moves something 1 ft (foot).

Energy is being used as effort to lift ice dancer

▶▶ **Read further › gravity**
pg22 (m12)

▶ *The lifting force of the ice dancer overcomes the force of gravity as he lifts the load.*

Ice dancer being lifted is affected by force of gravity

▶▶ **Read further › energy**
pg24 (d2)

Acceleration and mass

In the 17th century, English scientist Isaac Newton realized that forces work the same way everywhere in the universe, and that their effect can be predicted. A force makes objects accelerate. Just how much they accelerate depends on how strong the force is and how great the object's mass is – how much matter it contains. The bigger the force, the greater the acceleration. More massive objects need a bigger force to give them the same acceleration.

BRIGHT SPARKS

• Like all forces, gravity makes things accelerate. Anything falling towards the Earth accelerates, gaining speed at the rate of 9.8 m/sec, when near the Earth. This is called acceleration due to gravity, or g.

• Objects resist changes of motion. The more mass (matter) a substance contains, the more it resists changes. The resistance is called inertia.

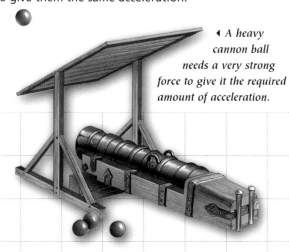

◀ *A heavy cannon ball needs a very strong force to give it the required amount of acceleration.*

The Laws of Motion

In the late 17th century, Isaac Newton summed up the link between force and motion with three laws: Law one: an object only accelerates (changes speed or direction) when a force acts on it. Law two: acceleration increases as the force increases, but decreases as the mass increases. Law three: every action (force) is matched by an equal and opposite reaction (opposing force). These three laws apply to everything from kicking a ball to flying a spacecraft.

▶▶ **Read further › Newton**
pg22 (m10)

▼ *Without gravity or air resistance a kicked football would continue on its path.*

Air resistance slows down the ball

Force is applied when the ball is kicked

Gravity pulls down the ball

Air resistance and gravity combine to bring the ball back down to the ground

Anything moving in a straight line at a consistent level of force, will go on for ever, without an opposite force to stop it

Energy and work

ENERGY IS the ability to make something happen. It is not just the light that comes from the Sun, or the heat that comes from a fire. Scientists say that energy is the capacity to do work. It is involved in everything that happens everywhere in the universe, however tiny or gigantic, from grass growing to stars exploding. All substances have energy locked up inside their atoms and molecules. Energy comes in many different forms and it can change from one form to another.

● BRIGHT SPARKS

• Energy can never be created or destroyed; it can only be changed from one form to another. So all the energy in the universe has always existed no matter what form it is in now.

• Scientists define work as a force multiplied by the distance a load moves when the force acts on it. In the metric system, the unit of energy is called the joule. One joule is the work done when a force of 1 newton moves an object a distance of 1 metre, or a metre-newton. The foot-pound is the imperial energy unit used in countries such as the USA.

● IT'S A FACT

• All the energy in the universe may eventually turn to heat. This is called the 'Heat Death of the Universe' theory.

• Walking needs about five times as much energy as sitting still; running needs about seven times more energy.

● Movement energy

A moving object has a type of energy called kinetic energy. The more massive the object is and the faster it moves, the more kinetic energy it has. When runners set off at the beginning of a race, they convert chemical energy in their muscles to kinetic energy. The faster they can change chemical energy into kinetic energy, the faster they run. At the end of the race they stop producing kinetic energy. Air resistance and friction between their shoes and the ground slow them down.

▶▶ Read further › chemical energy
pg14 (d2)

▼ For a split second as the sprinters come away from the starting blocks, they accelerate faster than an average sports car.

▸ At the end of the race, the runners quickly lose their kinetic energy and come to a halt.

Check it out!
• http://www.eia.doe.gov/kids/
• http://www.energyquest.ca.gov

A hot cup of coffee has about 40 per cent more energy than a tennis ball served at 25 m/sec

1   2   3   4   5   6   7   8   9   10   11   12   13   14   15   16   17   18   19

▼ Power stations are giant energy converters: they convert chemical or nuclear energy in a fuel into heat.

● Converting energy

The heat generated by power stations is transferred to water. This changes into steam, which is released out of large, concrete chimneys. The kinetic energy of the steam makes a turbine spin. The rotational (spinning) energy of the turbine drives a generator, which converts the rotational energy into electrical energy for us all to use.

▶▶ Read further › chemical energy
pg14 [d2]

▸ When sea waves crash onto a shore, the energy they carry can move sand and stones, or even make some cliffs collapse.

● Potential energy

Things can have energy because of where they are. This type of energy is called potential energy. It is stored energy. A crane has to work against gravity to lift an object high above the ground. The object stores the energy needed to lift it. If the crane drops it, its potential energy changes to motion as it hurtles towards the ground.

▶▶ Read further › gravity
pg22 [m12]

▼ When something is lifted, it stores the energy used to lift it as potential energy.

● Making waves

All waves, including water waves and electromagnetic waves, carry energy. When waves hit something they give up some or all of their energy. If a pebble is thrown into water, the vibration travels outwards as a wave. When light waves strike the back of our eyes their energy affects the retina (the light-sensitive part) and we see things. When infrared waves hit something their energy changes into heat. When radio waves hit the aerial of a radio set, their energy changes into electric currents that the radio set changes into sound.

▶▶ Read further › gravity / waves
pg22 [m12]; pg29 [o22]

A huddle of ten children will heat a room as effectively as a one-kilowatt fire

a b c d e f g h i j k l m n o p q r s t u v w

Heat

HEAT IS another name for 'internal energy' or energy that is stored inside a substance. It is a form of energy that moves from one place to another when its temperatures are different. You can give extra internal energy to a substance by heating it or doing work on it. A bicycle pump warms up when you use it because the air inside it is squashed every time you push the handle. The work put into squashing the air gives it more energy, which makes its atoms and molecules move around faster. Every time energy changes from one form to another, some of it turns into heat, which then spreads to its surroundings. That is why computers, television sets and all sorts of machines often heat up when they are working.

▼ A refrigerator is a heat pump – it pumps heat energy from cold to hot, opposite to the way it flows in nature.

● Expansion and contraction

When things heat up they expand because their molecules vibrate more and move further apart. When they cool down, they contract because the molecules vibrate less and move closer together. With solids, the expansion and contraction is often too small to see. For example, a steel bar becomes only 0.0001 per cent longer for every degree rise in temperature. But the force of expansion is so powerful that extreme heat can make, for example, railway tracks buckle or doors jam.

▶▶ **Read further > solids**
pg9 (b22)

Iron

Copper

Heating element

● Hot and cold

Heat always flows from hot to cold in nature, but it can be made to flow in the opposite direction by using a refrigerator. Heat from food inside the refrigerator warms a special liquid flowing through pipes in the fridge. The heat makes the liquid evaporate (change from a liquid to a gas). The gas is piped outside the refrigerator, where it is first cooled and then compressed (squashed) to change it back to a liquid. It goes back inside the refrigerator where it collects more heat from the food and evaporates.

▶▶ **Read further > changing states**
pg8 (d2)

◀ Bimetallic strips used in thermostats to respond to a change in temperature consist of iron and copper bonded firmly together. The strip bends when heated because the copper expands more than the iron.

▶▶ Read further › radiation
pg21 (b22)

● Moving heat

Heat spreads in three ways: conduction, convection and radiation. Conduction is the spread of heat from atom to atom. The atoms in a hot substance move fast and hit each other. These collisions pass energy on to nearby atoms, and those atoms pass their energy on to the next atoms, and so on. Convection spreads heat in gases (and liquids). When they are heated their molecules speed up, collide more often and spread out further apart. The gas (or liquid) becomes lighter than the surrounding, cooler gas (or liquid) and floats upwards. Convection currents carry warm air up from a radiator. Radiation spreads heat by means of invisible infrared (heat) rays.

◀ *Hot liquid heats a metal spoon by conduction.*

▲ *Radiation is one of the ways that heat spreads from a flame.*

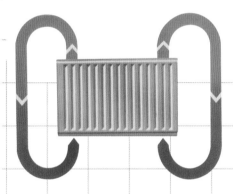

◀ *Air warmed by a radiator gets lighter and rises. Cold air moves in from below to replace it, and this is heated up too.*

▶▶ Read further › temperature
pg9 (b33)

• Temperature is measured in degrees on a temperature scale. The most common temperature scales are Celsius (Centigrade) and Fahrenheit. On the Celsius scale, the freezing point of water is 0°C and its boiling point is 100°C. On the Fahrenheit scale, these two points are at 32°F and 212°F.

• A hot-air balloon flies because the air trapped inside it is warmer than the surrounding air. The trapped air floats upwards, carrying the balloon with it.

▼ *Double glazed windows are used in most commerical buildings to slow down the loss of heat energy.*

● BRIGHT SPARKS

• Heat is not the same as temperature. Temperature is a measure of how fast molecules are moving. Heat is the movement energy of all the molecules added together.

• Temperature is measured by a thermometer. When a thermometer is dipped in something hot, it quickly reaches the same temperature. A liquid (usually mercury or alcohol are used) inside the thermometer expands as it heats up and spreads out along a channel inside the thermometer. The hotter the liquid, the more it spreads.

● Stopping the flow

Sometimes it is necessary to stop heat moving and keep it in one place. When a building is heated during winter the heat tries to flow out to make the building the same temperature as the air outside. Glass conducts heat faster than walls and roofs so a lot of heat escapes through windows. Therefore, some buildings are fitted with double-glazed windows to help keep the heat in. Double-glazed windows have two sheets of glass instead of one, with a small gap between them. The air in the gap is a poor conductor of heat.

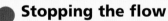

Light

LIGHT IS the only kind of electromagnetic radiation to which our eyes are sensitive. We are surrounded by light most of the time but surprisingly few things actually give out light. The Sun is our main source of light *(see pg20 [t5])*. Light also comes from stars, candles, electric lights and even small insects, such as glow-worms. But most things are only visible because they reflect light from other sources.

▶▶ Read further > waves
pg25 [m34]

Light and shade

When light rays hit an object, they bounce off, are absorbed or pass through. Substances, such as clear glass, which lets light through without breaking the direction of the rays, are transparent. Substances that jumble the light on the way through, such as frosted glass, are translucent. Those that stop the light altogether are opaque.

Casting a shadow

Light travels in straight lines. Different substances allow different amounts of light to pass through them. When light hits an opaque object, it does not bend around it. Opaque objects cast two types of shadow: if no light reaches an area, a dark shadow called an umbra is formed. If some light reaches an area, a grey shadow called a penumbra is formed.

▼ *The pillars at Stonehenge are opaque, so when sunlight falls on them, the area behind them receives no light and is left in shadow as light cannot pass through.*

Clear glass is transparent

Frosted glass is translucent

China objects are opaque

▶ *Not all light gives off heat. Glow-worms and fireflies attract a mate by making a part of their body glow using chemical reactions inside their bodies.*

Reflection and refraction

When light strikes a surface, some or all of it is reflected. From most surfaces it is scattered in all directions. On mirrors and other smooth shiny surfaces every ray may bounce off in exactly the same pattern as it arrives. The result is a perfect reflection of the light source or a mirror image. When light passes through something transparent, such as water, the rays are bent or refracted as the light slows down. This is why swimming pools sometimes look shallower than they really are, and straws appear bent when standing in a glass of water.

◀ *Refraction makes a straw look bent. It is not really bent: the light rays bend when they pass through the water.*

Read further › solar rays pg21 (n22)

Colours of light

When we see different colours, we are seeing different wavelengths of light. Sunlight appears colourless but it is called white light, which is actually a mixture of all colours. When it hits raindrops in the sky it can be split up into all the different colours to form a rainbow. They are split because the raindrop refracts (bends) each wavelength of light differently, fanning out the colours in a particular order, from red (the longest waves) at one end, to violet (the shortest) at the other.

Read further › Sun's rays pg21 (b22; n22)

◀ *The whole band of colour is called the spectrum. The individual colours are created by splitting white light.*

How light travels

Light is the fastest thing in the universe. In space, it travels at 299,792,458 m/sec and takes just eight minutes to reach Earth from the Sun. It travels slightly slower in air, and slower still in water, but still travels very fast. Light always travels in tiny waves. But these waves are tiny vibrations of energy, not a bit like ripples in a pond. Every beam of light is a stream of tiny packets of these vibrations called photons, each with its own wavelength. Photons are particles like electrons, but so small they have no mass at all.

BRIGHT SPARKS

• Fluorescent lights can be made from a tube filled with a gas called neon. When electricity passes through the tube, energy is given to particles in the gas, which transfer this into light – neon light.

• Scientists think the human eye is so sensitive that a single photon falling on the eye can trigger a signal to the brain.

▼ *The waves in the electromagnetic spectrum all differ in length, but all produce the same form of energy – just at different levels.*

Long radio waves   Microwaves   X-rays   Gamma rays

Shorter radio waves (TV)   Light waves   Short X-rays

Read further › light pg25 (m34)

Check it out!
• http://unmuseum.mus.pa.us/speed.htm

Scientists recently discovered that they can make objects travel faster than light

Sound

EVERY SOUND you hear, from a child's cry to the roar of an engine, is created by a vibration. Sometimes you can see the vibration – such as a guitar string twanging quickly back and forth. Often the movement is invisible, but the vibration is always there, and as the sound source moves to and fro it pushes the air around it to and fro as well. The air is alternately squeezed and stretched to create waves that spread out in all directions. When these sound waves reach you, the sensitive mechanisms in your ears respond to the air vibrations, and so you hear sound.

Waves of sound

Sound waves radiate in all directions from the sound source. Sound can travel through liquids, such as water, and many hard solids too. But a vacuum – completely empty space – is completely silent, because there is nothing to transmit the sound waves.

▶▶ Read further > solids / waves
pg8 (d2); pg25 (m34)

Echoes and acoustics

If you shout in a large empty hall you may hear the sound ringing out momentarily afterwards. This is because the sound of your voice bounces back to you off the hard walls. Every smooth, hard surface, such as a wall, bounces sound back, but you only hear echoes in big, empty spaces where the walls are far enough away for there to be a gap between the original sound and its return. Although you may not always hear a distinct echo, sound reflections always affect the quality of the sound you hear. Concert halls must be designed carefully so that the internal surfaces of the hall create the right acoustics – sound vibrations – for the musicians and orchestras playing.

▸ Whales and dolphins make high-pitched squeaks, clicks and whistles that bounce off the seabed and surrounding fish or rocks. The echo that returns to the whale or dolphin helps it to find food.

▶▶ Read further > reflection
pg29 (b22)

Check it out!

• http://www.physicsclassroom.com/Class/sound/soundtoc.html

Sounds too high-pitched for humans to hear are called ultrasounds

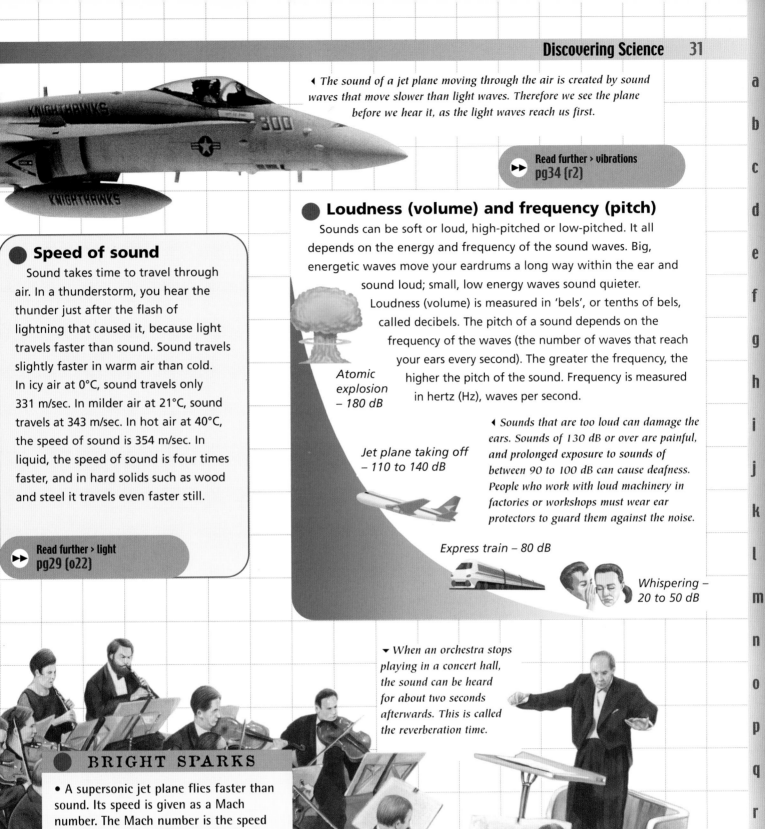

◀ *The sound of a jet plane moving through the air is created by sound waves that move slower than light waves. Therefore we see the plane before we hear it, as the light waves reach us first.*

▶▶ **Read further › vibrations**
pg34 (r2)

● Loudness (volume) and frequency (pitch)

Sounds can be soft or loud, high-pitched or low-pitched. It all depends on the energy and frequency of the sound waves. Big, energetic waves move your eardrums a long way within the ear and sound loud; small, low energy waves sound quieter.

Loudness (volume) is measured in 'bels', or tenths of bels, called decibels. The pitch of a sound depends on the frequency of the waves (the number of waves that reach your ears every second). The greater the frequency, the higher the pitch of the sound. Frequency is measured in hertz (Hz), waves per second.

● Speed of sound

Sound takes time to travel through air. In a thunderstorm, you hear the thunder just after the flash of lightning that caused it, because light travels faster than sound. Sound travels slightly faster in warm air than cold. In icy air at 0°C, sound travels only 331 m/sec. In milder air at 21°C, sound travels at 343 m/sec. In hot air at 40°C, the speed of sound is 354 m/sec. In liquid, the speed of sound is four times faster, and in hard solids such as wood and steel it travels even faster still.

▶▶ **Read further › light**
pg29 (o22)

Atomic explosion – 180 dB

Jet plane taking off – 110 to 140 dB

◀ *Sounds that are too loud can damage the ears. Sounds of 130 dB or over are painful, and prolonged exposure to sounds of between 90 to 100 dB can cause deafness. People who work with loud machinery in factories or workshops must wear ear protectors to guard them against the noise.*

Express train – 80 dB

Whispering – 20 to 50 dB

▼ *When an orchestra stops playing in a concert hall, the sound can be heard for about two seconds afterwards. This is called the reverberation time.*

BRIGHT SPARKS

• A supersonic jet plane flies faster than sound. Its speed is given as a Mach number. The Mach number is the speed of the plane divided by the speed of sound in the air the plane is flying through. Mach 1 is the speed of sound, Mach 2 is twice the speed of sound.

• Most sounds contain lots of different frequencies jumbled together – some are louder than others. A guitar and a violin playing the same note sound different, because they produce a different mix of strong and weak frequencies.

Dogs can hear sounds more than twice as high pitched as humans can

Air and water

A IR AND water are the two most important substances in the world. Without air and water, life on Earth would be impossible. Air not only provides living creatures with the oxygen they need to breath, but also a place to move around easily, too. The Earth's blanket of air, called the atmosphere, provides vital protection against the harmful radiations from space and helps to maintain a stable environment in which life can flourish.

▼ *Plants need water to grow and survive. Rainforests receive a lot of rain annually, and so are abundant with plant life and animals. In an ecosystem, each organism (living thing) is dependent on the other organisms living there.*

IT'S A FACT

• At the temperature called the triple point (0°C), water can exist in all three states (solid, liquid and gas) at the same time.

• Only 2 per cent of the world's water is permanently frozen in ice caps and glaciers.

BRIGHT SPARKS

• Water is less dense as a solid (ice) than as a liquid, which is why ice floats. Water actually expands as it freezes, which is why pipes burst in cold winters and frost can split rock.

• Carbon dioxide (CO_2) makes up 0.03 per cent of the air, but the proportion is continually changing as it is breathed out by animals and taken in by plants as they grow. Extra gas pumped out by industries and vehicles burning oil boosts CO_2 levels dramatically.

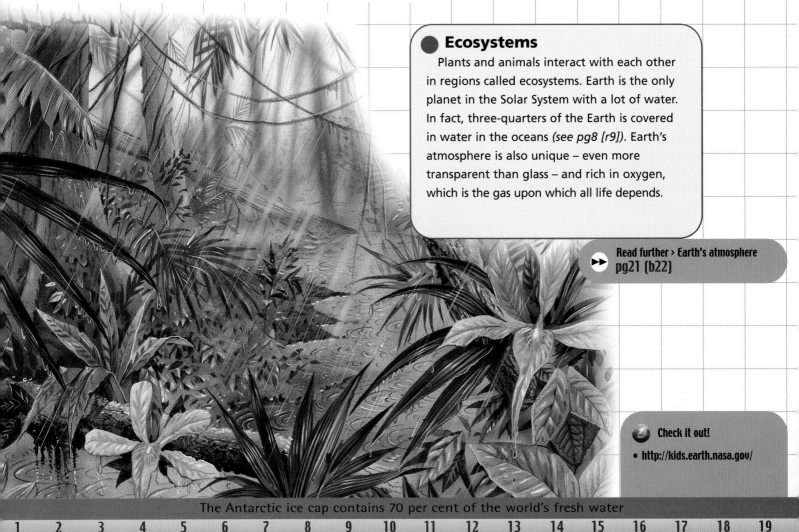

Ecosystems

Plants and animals interact with each other in regions called ecosystems. Earth is the only planet in the Solar System with a lot of water. In fact, three-quarters of the Earth is covered in water in the oceans *(see pg8 [r9])*. Earth's atmosphere is also unique – even more transparent than glass – and rich in oxygen, which is the gas upon which all life depends.

▶▶ Read further › Earth's atmosphere pg21 (b22)

Check it out!
• http://kids.earth.nasa.gov/

The Antarctic ice cap contains 70 per cent of the world's fresh water

a
b
c
d
e
f
g
h
i
j
k
l
m
n
o
p
q
r
s
t
u
v
w

● What are air and water?

Water is a compound in which two atoms of hydrogen gas (H_2) combine with one atom of oxygen gas (O) to form a colourless, odourless, tasteless liquid (H_2O). Sometimes, though, it may contain dissolved traces of other chemicals. Air, however, is always a mixture – not a compound – containing a variety of gases that are mixed but not chemically linked. More than three-quarters of the air is nitrogen (about 78 per cent), and most of the rest is oxygen (about 21 per cent). The remaining 1 per cent consists of carbon dioxide, water vapour and traces of other gases such as neon, helium, ozone and krypton.

Nitrogen
78.08%

Oxygen
20.94%

Carbon
dioxide
0.03%

Argon and
other gases
0.95%

▶▶ **Read further › gases**
pg9 (q31); pg13 (b29)

◀ *Air is a unique mixture that exists on Earth and nowhere else in the Solar System.*

▶▶ **Read further › attraction**
pg19 (b31)

● Why is water liquid?

Water is the only substance that can be solid, liquid or gas at everyday temperatures. As a water molecule consists of two hydrogen atoms and one oxygen atom, the hydrogen atoms at one end of the molecule have a positive electrical charge; the oxygen atom at the other end has a negative charge. Molecules like this are called polar molecules. The positive and negative charges attract each other and hold water molecules together. Without them the molecules would come apart from each other more easily and water would change to steam at a lower temperature.

● What is a solution?

A remarkable property of water is its ability to make solutions with other substances. A solution is formed when a substance is added to a liquid and instead of just floating in the liquid, the substance breaks up entirely, so that its atoms and molecules completely intermingle with those of the liquid. This happens when coffee is added to water. The substance dissolved is called the solute; the liquid is called the solvent.

▶ *When sugar dissolves in coffee, the sugar is the solute and the coffee is the solvent.*

▶▶ **Read further › crystals / salt**
pg15 (k31)

▲ *Ice is slightly less dense than water, which is why icebergs float in water, often partially above the surface depending on the volume (size) of the iceberg.*

Time

BEFORE THE invention of clocks, people measured time by the natural rhythm of the Earth – by watching the movements of the Sun, Moon and stars in the sky. Now we can see time passing as hands move or figures change on a clock. Modern atomic clocks can measure time with astonishing accuracy. Yet scientists and philosophers still find it hard to agree on what time is. Scientists say time is a dimension (like length and width) and that we move through time just as we can move sideways or backwards and forwards, or up and down. Time is said to be the fourth dimension: the other three are length, width and depth. But time only runs one way: a candle cannot be unburned, or your life lived backwards.

Detector counts the atoms

Magnet separates atoms

Microwave source

▶ *Atomic time is measured by how many electromagnetic waves are absorbed by atoms.*

Oven where atoms 'boil off'

Frequency divider

Computer adjusts microwaves

17:00:1070

Digital display to show time

● Atomic time

Just as a guitar string vibrates at a certain pitch or frequency, so do atoms and molecules. Atomic vibrations are so regular that they can be used to make the world's most accurate clocks – atomic clocks. These special clocks, all housed in special laboratories, mostly use the caesium-133 atom *(see pg12 [r2])*. Since 1967, 1 second has been defined as 9,192,631,770 vibrations of a caesium-133 atom. Atomic clocks are also used to set the world's standard time, called Co-ordinated Universal Time (UTC), set by the United States National Institute of Standards and Technology (NIST) agency.

▶▶ Read further > vibrations
pg30 (d2)

● Check it out!
• http://www.mrdowling.com/ 601-time.html

A solar day is exactly 24 hours, the time from midday one day to midday the next

| 1 | 2 | 3 | 4 | 5 | 6 | 7 | 8 | 9 | 10 | 11 | 12 | 13 | 14 | 15 | 16 | 17 | 18 | 19 |

Natural clock

Sundials work well in the sunshine, but they do not work at night or in cloudy weather. Other ways of measuring time that did not depend on sunshine were invented in the ancient world. A candle burns down steadily, so time can be measured by the changing length of a burning candle. Water or sand running steadily from one container to another can also mark the passing of time. Then in the 17th century, the great Italian scientist Galileo Galilei noticed that a pendulum of a certain length (a weight on the end of a wire or rod) always swings at the same rate. This discovery made it possible to build accurate clocks by linking the swinging pendulum to pointers (hands) that showed the time on a dial (clock face).

▲ *As the Sun moves from east at the beginning of the day to west at the end of the day, the position of the shadow cast on the sundial moves, showing what time of day it is.*

Read further › shadows
pg28 (k2)

BRIGHT SPARKS

• The length of a day on Earth has not always been the same. In the past, days were shorter and in the future, days will be longer because the Earth is spinning more slowly as time goes on. The average day is 1.6 milliseconds (thousandths of a second) longer every 100 years.

• Einstein showed that the faster things travel, the slower time runs. When *Apollo 11* went to the Moon, an accurate clock on board lost a few seconds, not because of any fault but because time ran slowly in the speeding space ship.

▼ *Time travel has been the subject of many books and films. The machine in H.G. Wells' classic novel* The Time Machine *is portrayed on screen as being able to withstand the immense pressures of wormholes.*

Time travel

The brilliant scientist Albert Einstein suggested that time does not exist on its own but is part of something called space-time, which includes the whole universe. When we move about, we are not moving through space while time passes separately. We move through space-time. If time is linked to space, some scientists wonder if we might be able to move through the time part of space-time into the past or future. Objects in space called black holes have such a strong force of gravity that they might be able to bend space-time enough to create pathways called worm-holes that link to other places or times. If wormholes do exist then they are understood to be even smaller than an atom. Some scientists believe it might be possible to enlarge them using an extremely powerful electric field, holding them open long enough to make a tunnel through space-time and thus travel through time.

Read further › gravity
pg22 (m12)

A day measured by the stars (a sidereal day) is 23 hours 56 minutes and 4.09 seconds

a b c d e f g h i j k l m n o p q r s t u v w

Glossary

Acceleration A change in speed or direction.

Acid A solution made when substances containing hydrogen dissolve in water. Some acids taste sour. Others are highly corrosive.

Alkali A base that dissolves in water.

Allotrope One of various forms in which an element occurs. For example, diamond and graphite are allotropes of carbon.

Alloy Metal mixed with another metal or substance. Steel is a tough alloy made from iron, carbon and traces of other substances.

Atom The smallest particle of an element.

Base The chemical opposite of an acid. Weak bases taste bitter and feel soapy. Strong bases are corrosive.

Carbon An element found as coal, diamond and various other substances. Compounds that it makes are called organic chemicals.

Compound Substances made by the atoms of two or more elements chemically bonding.

Conduction The spreading of heat from hot areas to cold areas by direct contact.

Contraction When something gets smaller, usually as it gets colder.

Convection When warm air or liquid rises through cool air or liquid.

Decibel A measure of the loudness of sound.

Effort The force to make something move.

Electrical charge The force of attraction between certain sub-atomic particles.

Electron The tiniest of three atomic particles that circle an atom's nucleus.

Element The simplest possible chemicals, made entirely of their own unique atom.

Energy The power to make things happen.

Evaporation When liquid turns to vapour as it gets warm.

Expansion When something gets bigger, usually as it gets warmer.

Force Something that changes an object's shape or the way it is moving, by pulling, pushing, stretching or squashing it.

Gas Substance that has no shape or form.

Heat The total energy a warm substance has because of the movement of its molecules.

Magnetism The force of attraction or repulsion in substances such as iron, and around a moving electrical charge or current.

Mass The amount of matter in an object.

Matter Anything of physical substance in the universe.

Melting The way a solid substance turns to liquid when it reaches a certain temperature.

Molecule The smallest bit of a substance that can exist by itself.

Momentum An object carries on moving at the same speed and in the same direction.

Neutrino One of the tiniest of all particles from which matter is made.

Neutron With protons, one of the two basic particles that make up an atom's nucleus.

Nuclear energy Energy released by splitting or fusing the nuclei of atoms.

Nucleus (plural nuclei) The core of an atom, made of proton and neutron particles.

Particle A very small piece of matter.

Periodic Table The table of chemical elements, arranged in order of the number of protons in the nucleus of their atoms.

Photon One of the tiny particles or 'packages' that light travels in.

Plasma Special form of gas at very high temperatures, full of charged particles.

Pole, Magnetic The region of a magnet where the magnetism is strongest. Every magnet has north and south poles.

Polymer Substance such as plastic made from long chain or branchlike molecules.

Proton With neutrons, one of the two basic particles that make up an atom's nucleus.

Radiation Energy emitted from particles as electromagnetic waves, such as light and radio waves, or as radioactive particles.

Reaction, Chemical The joining of two or more chemicals when at least one of the chemicals changes.

Reflection The bouncing back of light or other waves when they hit a surface.

Refraction Light rays bend as they pass from one transparent substance to another.

Solid State of matter with a shape and form.

Solution Liquid in which a solid, a liquid or a gas is broken up and intermingled so it behaves as part of the liquid.

Spectrum The complete range of colours contained in light.

Static electricity Electricity that builds up when things rub together but do not move.

Steam Gas formed by evaporating water.

Sub-atomic particles Particles that are even smaller than an atom.

Transition metal Metals including copper, cobalt and iron. Most are good conductors of electricity.

Volume The energy of a sound.

Wavelength The length from the crest of one wave to the crest of the next.

Index

Entries in bold refer to main subjects; entries in italics refer to illustrations.

A

absolute zero 8
acceleration 23, 24
acids 14, 15
acoustics 30, *30*
action 22
aerials 25
air 32–33
  carbon cycle 17
  convection 27
  resistance 17, 22, 23, *23*, 24
  sound waves 30, *30*, 31
  speed of light 29
alkali metals 12
alkaline-earth metals 12
allotropes 16
alloys 14, *14*
alternating current (AC) 19
aluminium 14, 15
amino acids 15
atmosphere 21, 32
atomic clocks 34, *34*
atomic explosions 31
atoms 10–11, 26
  carbon 16, 17
  conduction 27
  electricity 18
  elements 14
  energy 24
  metals 15
  radiation and 21
  solutions 33
  states of matter 8

B

bases 15, *15*
batteries 19
bimetallic strips 26, *26*
black holes 35
boiling 9

C

cables, electricity 18, *18*
calcium carbonate 15
carbon 16–17
  alloys 14
  carbon chemicals **16–17**
  carbon cycle 17, *17*
  carbon fibres 16
  diamonds 11
  organic compounds 17
  plastics 17
  symbol 13

carbon dioxide 10, 17, 32, 33
cars, metals 14, 15, *15*, 16, 17
caustic soda 15, *15*
cells, DNA 15
cellulose 17
Celsius scale 27
Centigrade scale 27
charcoal 16, *16*
charged particles 8, 19
chemical energy 24
chemicals 11, **14–15**
circuits, electric 19
citric acid 13
clocks 34, 35
colours 20, 29, *29*
  and heat 26
  primary colours 28
  rainbows 28, 29, *29*
  white light 29
compounds 13, *13*
  carbon 16
  metals 14
  organic chemistry 17
  water 33
compression, gases 26
computers 18, 26
conduction 27, *27*
conductors
  electricity 18, 19
  heat 15
contraction 26, *26*
convection 27, *27*
Co-ordinated Universal Time (UTC) 34
copper
  alloys 14
  bimetallic strips 26, *26*
  electricity 18, 19
  Periodic Table 12, *12*
crystals 11, 15
currents
  convection 27, *27*
  electricity 19

D

decibels 31, *31*
diamonds 11, *11*, 16, *16*
diesel oil 16, *16*
dimensions, time 34
direct current (DC) 9
DNA 15

E

Earth
  air and water 32, *32*, 33, *33*
  day length 35

Earth (*continued*)
  gravity 22, *22*, 23
  magnetic field 18, *18*
  radiation from Sun 21, *21*
  time 34, *34*
echoes 30
ecosystems 32, *32*
effort 23, *23*
electrical cables 18, *18*
electrical charges, particles 10, 19, *19*
electrical forces 22
electricity **18–19**
  conductors 15
  currents 18, 19
  fluorescent lights 29
  power stations 25
  static electricity 19
electromagnetic spectrum 29, *29*
electromagnetic waves 25, 28, 34
electromagnetism 18, 19, *19*, **20–21**
electrons 10, *10*, 11
  carbon 16
  electricity 18, 19
  metals 15
  Periodic Table 12, 13
  radiation and 21
elements
  atoms 11, 14
  Periodic Table 12–13
energy **24–25**
  atoms 10, *10*
  electricity **18–19**
  electromagnetism **20–21**, 29
  heat **26–27**
  nuclear energy 11
  states of matter 8, *8*
  Sun 21
environment 32
evaporation 26
expansion 26

F

fabrics 17
Fahrenheit scale 27
fields, magnetic 18, 19, *19*
fluorescent lights 29
foot-pounds 23, 24
forces **22–23**, *23*
freezing 8, 9, 14, 27, 32
frequency, sound 31, *31*
friction 24
fuels 20, 25
fullerenes 16, 17

G

gamma rays 20, *20*, 21, 29
gases **8–9**
  air 32–33
  atoms 10
  convection 27
  heat 26
  noble gases 13
  triple point 32
gems 11
generators 25
glaciers 32
glass 20, 27, 28, *28*
glucose 17
gold 12, 14, 15, 19
graphite 16, *16*
gravity 22, *22*
  black holes 35
  laws of motion 23
  parachutes 17
  potential energy 25

H

heat **26–27**
  bimetallic strips 26, *26*
  conductors 15
  energy 24
  expansion and contraction 26
  infrared waves 25
  movement 27
  radiation 20
helium 8, 9, 10, *10*, 13, 33
hertz (Hz) 31
hydrochloric acid 15
hydrogen
  acids 15
  atoms 10, *10*
  carbon compounds 17
  Periodic Table 12, 13
  protons 12
  water 14, 33

I

ice 8, 9, *9*
  density 33
  ice caps and glaciers 32
  icebergs 33, *33*
  melting point 9
inertia 23
infrared radiation 20. *20*, 25, 27
insulators 18
internal energy 26
ionization 21
ions 15
iron 12, 15
  alloys 14

iron (continued)
bimetallic strips 26, 26
magnets 19

J
joules 23, 24

K
kilograms 23
kinetic energy 24, 24, 25
krypton 13, 33

L
laws of motion 23, 23
lifting 25, 25
light 28–29
colours 28, 29, 29
Earth's atmosphere 21
eyesight 25
length 28
light waves 20
photons 22, 29
reflection and refraction
29, 29
shadows 28, 28
speed of 20, 29, 31
light bulbs 13, 13, 28
lightning 18, 31
limescale 15
limestone 16
liquids 8–9, 8, 9
convection 27, 27
evaporation 26
solutions 33
sound waves in 31
triple point 32
load 23, 24
loudness 31

M
Mach numbers 31
machines
heat 26
motion 23
sounds 30, 31
magnesium 14
magnetic attraction 19, 19
magnetism 18–19, 22
mass
acceleration 23
gravity 22
inertia 23
materials 14–15
matter
atoms 10–11
states of 8–9
melting 9, 9
mercury 14, 27

metals
alloys 14, 14
compounds 14
crystals 11
magnets 19
melting points 8
ores 14
Periodic Table 12, 12
metric system 23, 24
microscopes 10
microwaves 20, 20, 29
minerals 15
mixtures 14, 33
molecules 10–11
atoms 10–11
convection 27
crystals 11
energy 24
expansion and contraction
26
organic compounds 17, 17
solutions 33, 33
states of matter 8–9, 8–9
temperature 27
water 33, 33
momentum 22, 22
motion 22–23, 23, 25

N
neon 13, 29, 33
neutrinos 10
neutrons 10, 10, 11
newtons 23, 24
nitrogen 13, 33
noble gases 13
noise 31
non-metals 13
nuclear energy 11, 25, 25
nuclear forces 22
nucleus, atoms 10, 10, 11, 11,
12

O
oil 16, 20
opaque objects 28, 28
ores 14
organic chemistry 17, 17
oxygen 8
air 32, 33
atoms 10, 10
protons 12
water 14, 33
ozone 33

P
particles
charged 8, 19
photons 29

particles (continued)
sub-atomic 10
pendulums 35
penumbra 28
Periodic Table 12–13, 12–13
petroleum 16
phosphorus 13
photons 22, 28, 29
pico-seconds 34
pitch, sound 31, 31
plants
carbon cycle 17
ecosystems 32
fuels 20
plasma 8
plastics 16, 17, 17, 18
plutonium 12, 13
polar molecules 33
poles, magnets 19
polymers 17
potential energy 24, 24, 25
pounds 23
power stations 18, 19, 19,
25, 25
primary colours 28
proteins 15, 17
protons 10, 10, 11, 12
pylons, electricity 18, 18

R
radiation 20–21
Earth's atmosphere and
32
electromagnetic 20–21
heat 27
radiators 27
radio waves 20, 20, 21, 25,
29
radium 12
rainbows 20, 28, 29
rainforests 32
rare earths 12
reaction 22
reflection
light 29
sound 30
refraction 29, 29
resistance 22, 23, 24
reverberation time 31
rocks 8
atoms 10
compounds 14
crystals 11
salts 15

S
salt 11, 13, 15, 15
satellites 20, 21, 21

shadows 28, 28, 35
sodium 13
sodium chloride 13
sodium hydroxide 15
soil 14
solar days 34
Solar System 32, 33
solids 8–9, 8–9
crystals 11
expansion and contraction
26
melting 9
sound waves 31
triple point 32
solute 33
solutions 33, 33
solvents 33
sound 30–31
sound waves 21, 25,
30–31, 30
speed of 31
space, electromagnetic
waves 21
space-time 35
spacecraft
acceleration 22
laws of motion 23
radio waves 21, 21
time in 35
spectrum
colours 29
electromagnetic 20, 29
speed
acceleration 22, 23, 24
of light 20, 29, 31
of sound 31
stars
day length 35
energy 24
light waves 21, 28
time 34
states of matter 8–9
static electricity 19
steam 8, 9, 25, 33
steel
alloys 14
expansion 26
magnets 19
sound waves 31
uses 15
sub-atomic particles 10
substances 12–13
sugar
carbon cycle 17
crystals 11
solutions 33
sulphur 13
sulphuric acid 15

Sun
  light 24, *24*, 28
  ozone 33
  radiation 20, 21, *21*, 26
  temperature 8
  time 34, *34*

T
temperature 27
  heat 26
  lowest 8
  measurement 27
  states of matter 8, 9
  thermostats 26
thermal cameras 21, *21*
thermometers 27
thermostats 26
time 34–35
time travel 35, *35*

transition metals 12
translucent objects 28, *28*
transparent objects 28, 29, *29*
triple point, temperature 32
tungsten 8

u
ultraviolet radiation 20, *20*,
  21, *21*
universe
  atoms 10, 11
  electromagnetism 18
  energy 24
  gravity 22
  matter 10
  space-time 35
uranium 11, *11*, 13

V
vacuums 30
Van de Graaf generator 19, *19*
vibrations
  atomic clocks 34
  heat 26
  sound 30
  waves 25
volume, sound 31

W
water 9, *9*, 17, 32–33
  acids 14, 15
  atoms 10
  clocks 35
  light refraction 29
  molecules 33
  power stations 25
  sea water 14, 15

solutions 33
sound waves 30
speed of light 29
states of matter 8, 9
tap water 14, 15
triple point 32
water vapour 9, 33
waves
  electromagnetic **20–21**
  energy 25
  sea waves 25, *25*
  sound waves **30–31**
weather satellites 20
white light 29
wires, electricity 18, 19

X
X-rays 20, *20*, 21, 29

The publishers would like to thank the
following artists who have contributed to this book:
Mark Bergin, Syd Brak, Steve Caldwell, Kuo Kang Chen, Chris Forsey
Luigi Galante, Roger Goode, Janos Marffy, Helen Parsley
Martin Sanders, Rudi Vizi, Tony Wilkins

The publishers wish to thank the following sources for the photographs used in this book:
p35 (b/r) DreamWorks Pictures

All other photographs are from:
Corbis, Corel, digitalSTOCK, Flat Earth, Hemera,
MKP Archives, PhotoAlto, PhotoDisc

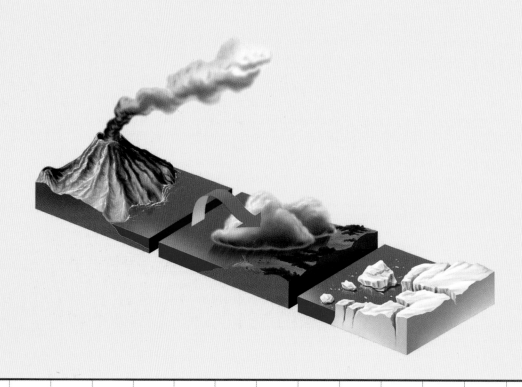